Birds, Bees and Burgers

Paulo Ferro and EnigMaths

 Tarquin

About the Authors

Paulo Ferro is a mathematics author, teacher and tutor from Coimbra in Portugal. He is well known around the world as EnigMaths and posts puzzles regularly on Instagram, Facebook and Twitter as enigmaths. He has taught and tutored mathematics for over 20 years. Published works include puzzles in Cambridge University's Plus Magazine, the NCTM website and the New York Times as well as academic publishing with Pearson, OUP, Editions Belin and Grupo Edebé.

Dedication

To my mom Lou for all her lifelong love and support.

ISBN (book): 978-1-91356-558-6
ISBN (ebook): 978-1-91356-559-6

Printed and designed in the UK

Published by Tarquin
Suite 74, 17 Holywell Hill
St Albans AL1 1DT
United Kingdom

info@tarquingroup.com
www.tarquingroup.com

Introduction

I hope that you enjoy these puzzles. I have chosen to present them in chapters labelled as Levels 1–6. Level 1 requires basic geometric knowledge – Level 6 requires more advanced geometric knowledge. 10 puzzles are presented in each chapter and then solutions for each are presented at the back of each chapter.

I hope you find the puzzles fun and challenging, and the solutions enlightening.

Contents

Level 1

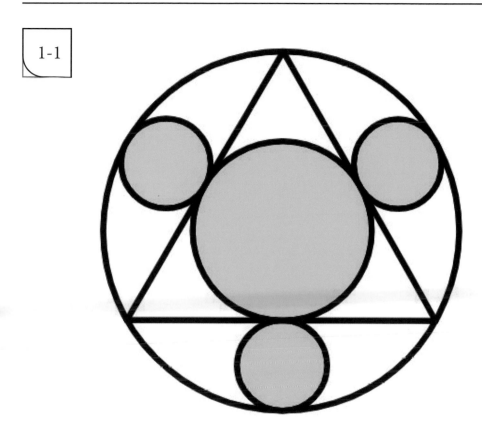

1-1

The equilateral triangle is inscribed in a circle.

What fraction of the big circle is shaded?

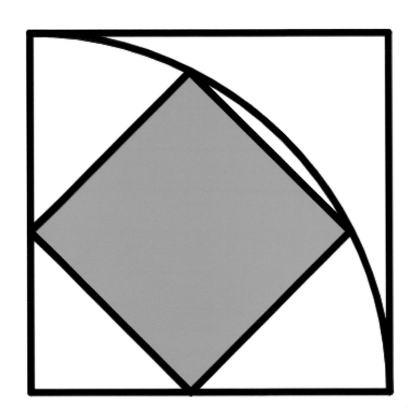

The big square has side with length 5.

What is the area of the small square?

1-3

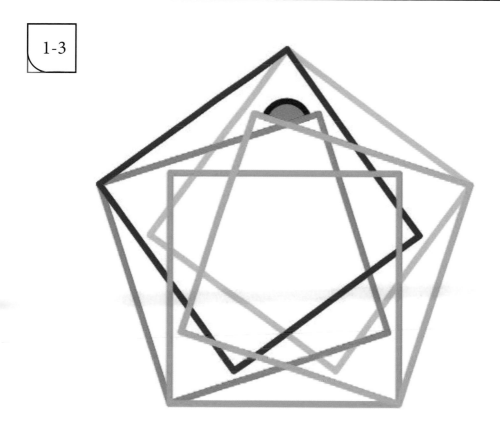

A regular pentagon is made with 5 congruent squares.

Find the missing angle.

1-4

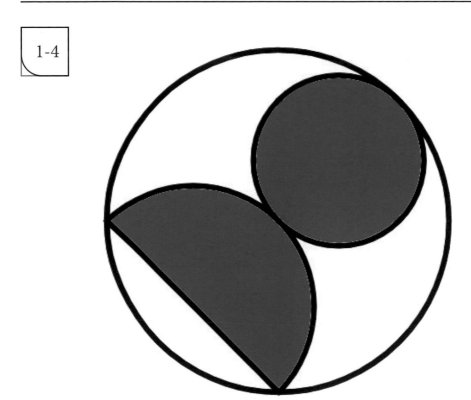

A circle and a semicircle meet at the centre of a big circle with radius 4.

What is the total shaded area?

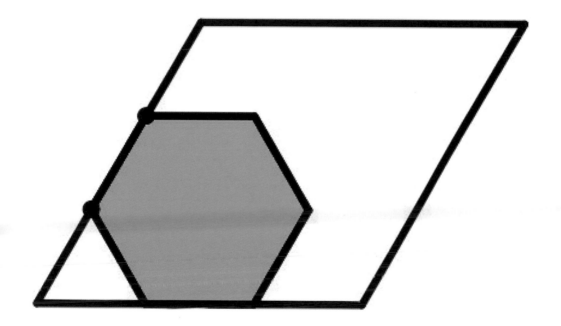

A regular hexagon is inside a rhombus such that two of its sides lie on the sides of the rhombus. The dots separate the side in 3 equal parts.

What fraction of the rhombus is shaded?

1-6

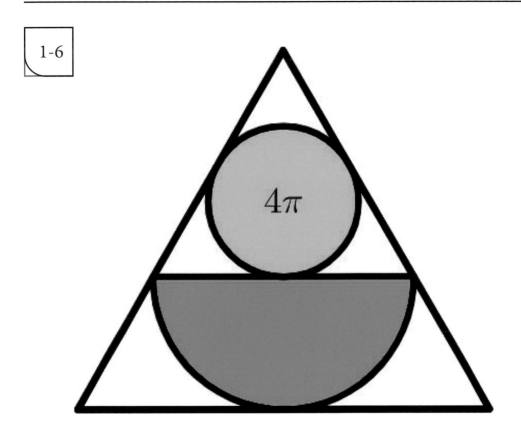

A circle and a semicircle are inside an equilateral triangle.

What is the area of the semicircle?

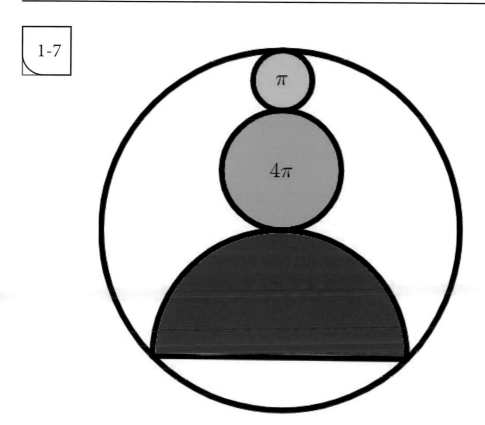

Two circles and a semicircle are inside a big circle. The second circle meet the semicircle at the centre of the big circle.

What is the area of the semicircle?

1-8

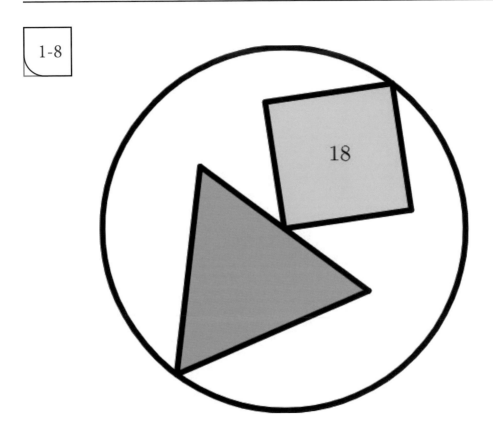

An equilateral triangle and a square meet at the centre of a circle.

What is the area of the equilateral triangle?

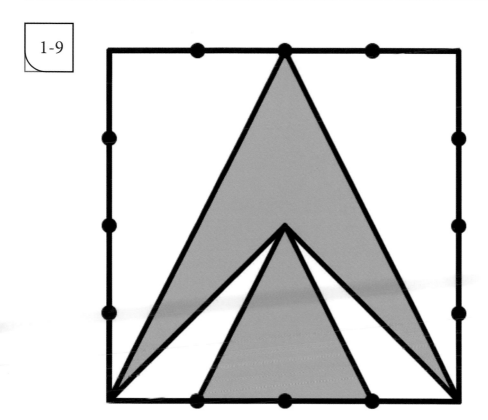

The points in each side are equally spaced.

What fraction of the square is shaded?

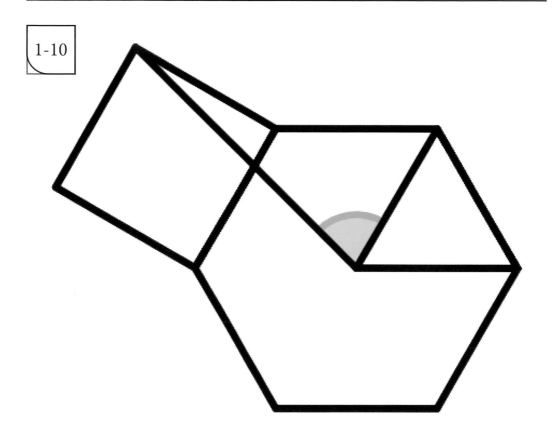

1-10

An equilateral triangle, a square and a regular hexagon have the same side length.

What is the size of the angle?

Solutions and Answers

1-1 Answer: $\dfrac{7}{16}$

Let r be the radius of the circle inscribed in the equilateral triangle. The circle where the equilateral triangle is inscribed has radius $2r$.

Each of the three small circles have diameter $2r - r = r$. So, their radii are $\dfrac{r}{2}$.

The area of the circle inscribed in the equilateral triangle is πr^2.

The area of each of the small circles is $\pi \left(\dfrac{r}{2}\right)^2 = \dfrac{\pi}{4}r^2$.

Then, the area shaded is $\dfrac{3\pi}{4}r^2 + \pi r^2 = \dfrac{7\pi}{4}r^2$.

The area of the circle where the equilateral triangle is inscribed is $\pi (2r)^2 = 4\pi r^2$.

So, the fraction shaded is $\dfrac{\dfrac{7\pi}{4}r^2}{4\pi r^2} = \dfrac{7}{16}$.

1-2 Answer: 10

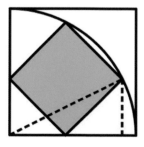

Let l_1 be the length of the side of the big square and l_2 the length of the side of the small square.

So, the diagonal of the small square is $\sqrt{2}l_2$.

Using the Pythagoras' theorem in the dashed right triangle to the left:

$$l_1^2 = \left(\frac{\sqrt{2}}{2}l_2\right)^2 + \left(\sqrt{2}l_2\right)^2$$

$$l_1^2 = \frac{1}{2}l_2^2 + 2l_2^2$$

$$l_1^2 = \frac{5}{2}l_2^2$$

$$l_2^2 = \frac{2}{5}l_1^2$$

$$l_2^2 = \frac{2}{5}(5)^2 = 10$$

So, the area of the small square is 10.

1-3 Answer: 144°

The shaded angle has the same size of the angle that belongs to the isosceles triangle in the bottom of the regular pentagon.

As the interior angles of a regular pentagon have size 108° and the interior angles of a square have size 90°, the isosceles triangle in the bottom of the regular pentagon has two equal angles with size $108° - 90° = 18°$.

Then, the missing angle measures $180° - 2 \times 18° = 144°$.

1-4 Answer: 8π

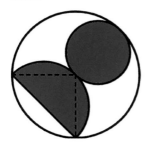

The radius of the small circle is half of the big circle, i.e., 2. Then, the area of the small circle is $(2)^2 \pi = 4\pi$.

The isosceles triangle inscribed in the semicircle has two sides (dashed line segments) with length 4, the radius of the big circle.

The diameter of the semicircle is $4\sqrt{2}$.

Then, the radius is $\dfrac{4\sqrt{2}}{2} = 2\sqrt{2}$.

So, the area of the semicircle is $\dfrac{\left(2\sqrt{2}\right)^2 \pi}{2} = \dfrac{8\pi}{2} = 4\pi$.

So, the shaded area is $2 \times 4\pi = 8\pi$.

1-5 Answer: $\dfrac{1}{3}$

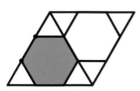

As it is a rhombus, all its sides have the same length. If we draw the diagonal of the rhombus that contains one of the sides of the regular hexagon and then reflects it about this diagonal we get the diagram to the left:

We have 2 congruent regular hexagons and 6 equilateral triangles inside the rhombus. These 6 equilateral triangles form 1 more regular hexagon with the same area of the other two.

So, the fraction of the rhombus shaded is $\dfrac{1}{3}$.

1-6 Answer: 6π

The circle is inscribed in a smaller equilateral triangle. Its radius is 2.
Let R be the radius of the semicircle.
Then,

$$(2+2)^2 = 2^2 + R^2$$
$$16 = 4 + R^2$$
$$R^2 = 12$$

So, the area of the semicircle is $\dfrac{12\pi}{2} = 6\pi$.

1-7 Answer: 9π

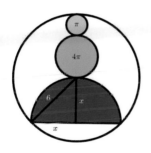

The radius of the first and second circles are 1 and 2, respectively.
So, the radius of the big circle is $2\times1+2\times2=2+4=6$.
We can draw a right isosceles triangle where the hypotenuse is the radius of the big circle.
Then,

$$6^2 = x^2 + x^2$$
$$36 = 2x^2$$
$$x^2 = 18$$
$$x = \sqrt{18}$$

So, the area of the semicircle is $\dfrac{\pi\left(\sqrt{18}\right)^2}{2} = \dfrac{18\pi}{2} = 9\pi$.

1-8 Answer: $12\sqrt{3}$

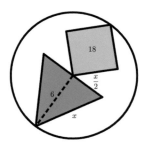

The radius of the circle is the diagonal of the square, i.e., $\sqrt{18}$.

So, the height of the equilateral triangle is $\sqrt{18}$.

Let x be the side of the equilateral triangle.

Using the Pythagoras' theorem to find the side of the equilateral triangle:

$$x^2 = 6^2 + \left(\frac{x}{2}\right)^2$$

$$x^2 - \frac{x^2}{4} = 36$$

$$\frac{3x^2}{4} = 36$$

$$x^2 = 48$$

$$x = \sqrt{48}$$

$$x = 4\sqrt{3}$$

So, the area of the equilateral triangle is $\dfrac{6 \times 4\sqrt{3}}{2} = 12\sqrt{3}$.

1-9 Answer: $\frac{3}{8}$

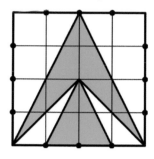

We can divide the square into 16 small squares:
The arrow can be divided into two obtuse triangles with the same base and the same height. Then, they have the same area, i.e., 2.

Then, the sum of their areas is $2+2=4$.

The isosceles triangle has area 2.

So, the fraction of the square shaded is

$$\frac{4+2}{16}$$
$$\frac{6}{16}$$
$$\frac{3}{8}$$

1-10 Answer: 75°

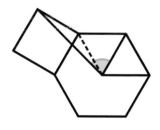

We can draw a line segment that creates another equilateral triangle and an isosceles triangle:

Let's concentrate our attention in the isosceles triangle. Its biggest angle has size $90° + 60° = 150°$.

Then the two equal angles have size $180° - 150° = 30°$, i.e., each of its equal angles has size $\frac{30°}{2} = 15°$.

So, the asked angle has size $60° + 15° = 75°$.

Level 2

2-1

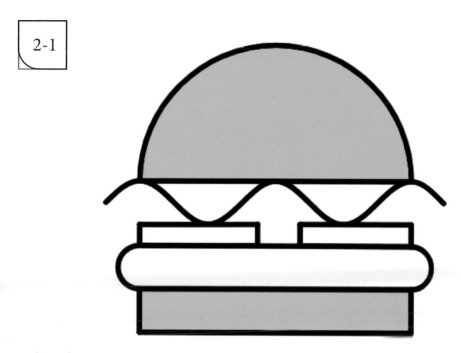

A hamburger is composed by lettuce, tomatocs and a patty of ground meat/mince placed inside a sliced bread roll.

The top slice of bread is represented by a semicircle, the lettuce by a part of the graph of $y = \sin x$ and the height of the bottom slice of the bread is equal to the height of the lettuce.

Find the area of the bread roll.

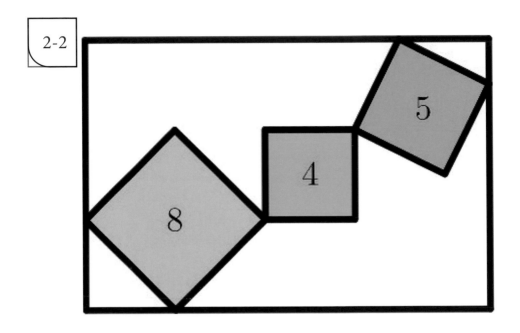

Three squares with areas 8, 4 and 5 are inside a rectangle.

What is the area of the rectangle?

2-3

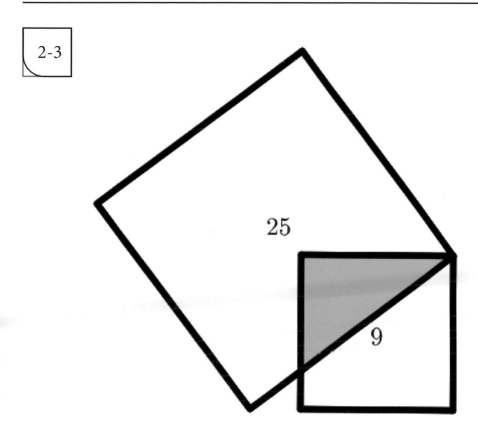

Two squares with area 9 and 25 intersect as shown in the diagram.

What fraction of the small square is shaded?

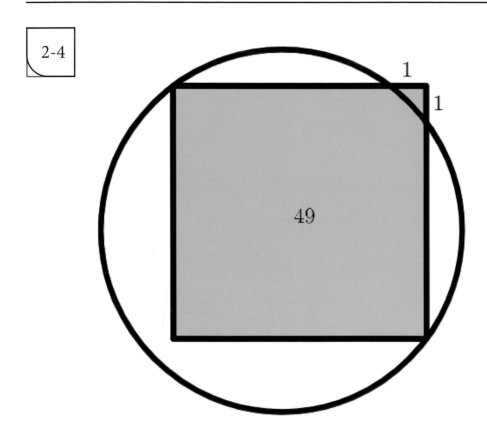

The square has area 49.

What is the area of the circle?

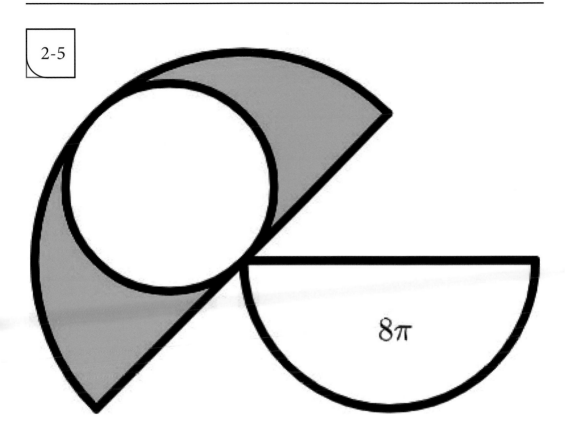

The small semicircle has area 8π. The big semicircle forms an angle of 45º with the small semicircle.

What is the shaded area?

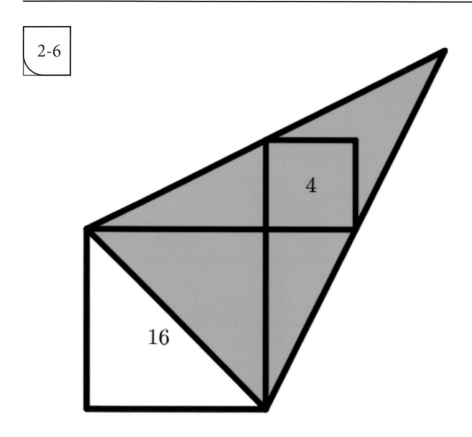

The big and the small squares have area 16 and 4, respectively.

What is the area of the shaded triangle?

2-7

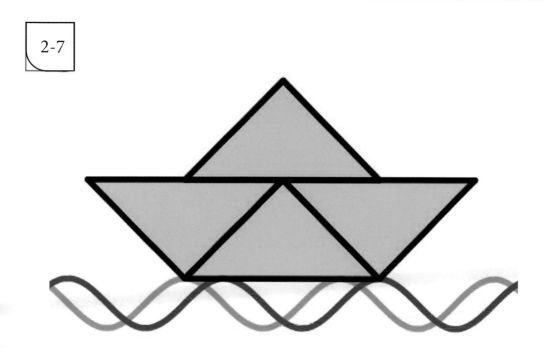

A boat with height 8, represented by 4 congruent isosceles triangles, is floating in the sea represented by a part of the graphs of $y = \sin x$ and $y = \cos x$.

What is the area of the boat?

A Ladybird/Ladybug has a form of a circle when its wing covers are closed. In a given moment, it opens them such that forms an equilateral triangle with side length $4\sqrt{3}$. The semicircles and circles in the wing covers have radius 1.

What is the area of the blue part of its wing covers?

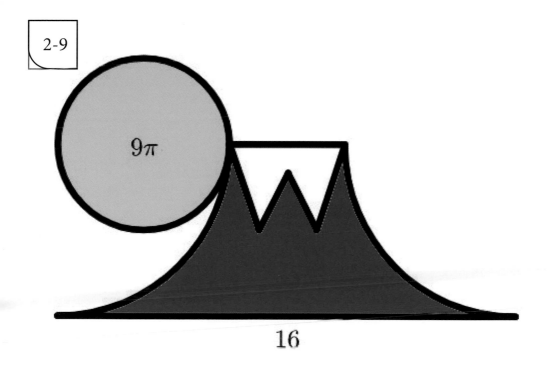

The Moon and Mount Fuji with snow on its top are represented above.

What is the area of the image of Mount Fuji, including the snow?

2-10

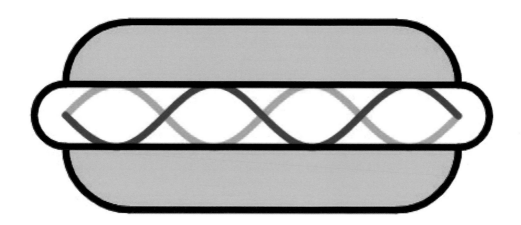

The diagram represents a hot dog. The ketchup and mustard are represented by the functions $y = \sin x$ and $y = \sin\left(x + \dfrac{\pi}{2}\right)$.

What is the area of the sausage?

Solutions and Answers

2-1 Answer: $2\pi^3 + 8\pi = 2\pi\left(\pi^2 + 4\right)$

The sine function and the diameter of the semicircle have three common points, in an interval of 4π (two times the period of the sine function).
So, the radius of the semicircle is 2π.

Then, the area of the semicircle is $\dfrac{\pi\left(2\pi\right)^2}{2} = 2\pi^3$.

The range of the sine function is between -1 and 1. Then, the height of the rectangle is 2 and the its area is $2 \times 4\pi = 8\pi$.

So, the area of the semicircle and the rectangle is $2\pi^3 + 8\pi = 2\pi\left(\pi^2 + 4\right)$.

2-2 Answer: 54

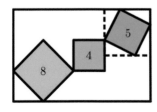

For the length:
The diagonal of the square in the left is

$$d^2 = \left(\sqrt{8}\right)^2 + \left(\sqrt{8}\right)^2$$
$$d^2 = 8 + 8$$
$$d^2 = 16$$
$$d = 4$$

The side of the square in the middle is 2.

The square in the right is inscribed in a bigger square of side length 3, forming four congruent right triangles with sides with lengths 1 and 2 (their hypotenuses are $\sqrt{5}$).

Then, the length of the rectangle is $4+2+3=9$.

For the width:

The left square has the diagonal with length 4, as we already saw.

The right square has the vertical side of the top right triangle equal to 2.

Then, the width of the rectangle is $4+2=6$.

So, the area of the rectangle is $9\times6=54$.

2-3 Answer: $\dfrac{3}{8}$

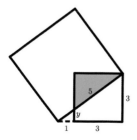

Using the Pythagoras' theorem in the dark blue right triangle:

$$5^2 = x^2 + 3^2$$
$$x^2 = 25 - 9$$
$$x^2 = 16$$
$$x = 4$$

Now, using the Thales's theorem in the same triangle:

$$\frac{y}{1} = \frac{3}{4}$$
$$y = \frac{3}{4}$$

So, one of the sides of the shaded triangle is 3 and the other is

$$3 - \frac{3}{4} = \frac{9}{4}.$$

The area of the shaded triangle is $\dfrac{\dfrac{9}{4} \times 3}{2} = \dfrac{27}{8}$.

Then, the fraction of the small square shaded is $\dfrac{\dfrac{27}{8}}{9} = \dfrac{3}{8}$.

2-4 Answer: 25π

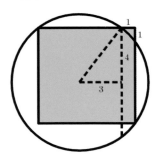

As the area of the square is 49, its side length is 7.

We can trace a vertical dashed line like shown in the diagram to the left:

Then we get a triangle where the horizontal side has length the half of 6, i.e., 3 and the vertical side has length the half of 8, i.e., 4.

Using the Pythagoras' theorem:

$$r^2 = 3^2 + 4^2$$
$$r^2 = 9 + 16$$
$$r^2 = 25$$
$$r = 5$$

So, the area of the circle is $\pi (5)^2 = 25\pi$.

2-5 Answer: 8π

The radius of the small semicircle is

$$\frac{\pi r^2}{2} = 8\pi$$
$$\pi r^2 = 16\pi$$
$$r^2 = 16$$
$$r = 4$$

We can draw an isosceles right triangle in the diagram:
The radius of the big semicircle is

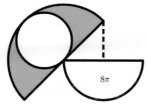

$$R^2 = 4^2 + 4^2$$
$$R^2 = 16 + 16$$
$$R^2 = 32$$
$$R = 4\sqrt{2}$$

Then, the area of the big semicircle is $\dfrac{\pi\left(4\sqrt{2}\right)^2}{2} = \dfrac{32}{2}\pi = 16\pi.$

The circle has diameter $4\sqrt{2}$. Then, its radius is $2\sqrt{2}$.

The area of the circle is $\pi\left(2\sqrt{2}\right)^2 = 8\pi.$

So, the shaded area is $16\pi - 8\pi = 8\pi.$

2-6 Answer: 24

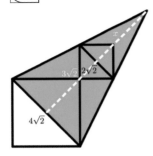

Sketching a diagonal of the small square and the height of the isosceles triangle in the diagram, we get two similar triangles: Then,

$$\frac{x}{2\sqrt{2}} = \frac{x+3\sqrt{2}}{4\sqrt{2}}$$

$$x = \frac{x+3\sqrt{2}}{2}$$

$$2x = x+3\sqrt{2}$$

$$x = 3\sqrt{2}$$

So, the area of the shaded triangle is $\dfrac{2\times3\sqrt{2}\times4\sqrt{2}}{2} = \dfrac{48}{2} = 24.$

2-7 Answer: 20π

As the periods of cosine and sine are 2π and they have their maximums (1) at 0 and $\dfrac{\pi}{2}$, respectively, the length of the bottom of the boat is $2\pi + \dfrac{\pi}{2} = \dfrac{5\pi}{2}.$

As the height of the boat is 8, the height of each triangle is 4.

Then, the area of one triangle is $\dfrac{4\times\dfrac{5\pi}{2}}{2} = 5\pi.$

So, the area of the boat is $4\times5\pi = 20\pi.$

2-8 | Answer: 13π

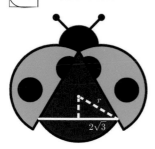

We can draw a right triangle using the centre of the circle. Then,

$$\sin 60° = \frac{2\sqrt{3}}{r}$$

$$\frac{\sqrt{3}}{2} = \frac{2\sqrt{3}}{r}$$

$$r = 4$$

So, the area of the wing covers is $\pi(4)^2 = 16\pi$.

As the radius of each small circles is 1, the area of the three circles is 3π.

So, the area of the blue part of the wing covers is $16\pi - 3\pi = 13\pi$.

2-9 | Answer: $6(16 - 3\pi)$

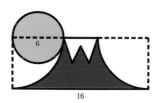

The radius of the Moon is 3 Then, the area of the quarter of the circle with radius $2 \times 3 = 6$ is $\dfrac{(6)^2 \pi}{4} = \dfrac{36\pi}{4} = 9\pi$.

The Mount Fuji image is a rectangle without two quarters of circle with the same radius.

So, the area of the Mount Fuji is

$$16 \times 6 - 2 \times 9\pi$$

$$96 - 18\pi$$

$$6(16 - 3\pi)$$

2-10 Answer: 9π

The sausage can be divided in 1 rectangle and 2 semicircles, one in each extremity. Both functions, $y = \sin x$ and $y = \sin\left(x + \dfrac{\pi}{2}\right)$, have period 2π. As there are 2 periods, the length of the rectangle is $2 \times 2\pi = 4\pi$.

The range of both functions is between -1 and 1. Then, the width of the rectangle is $1 - (-1) = 1 + 1 = 2$.

The radius of the 2 semicircles is 1.

So, the area of the sausage is

$$2 \times 4\pi + 2 \times \frac{\pi(1)^2}{2}$$

$$8\pi + \pi$$

$$9\pi$$

Level 3

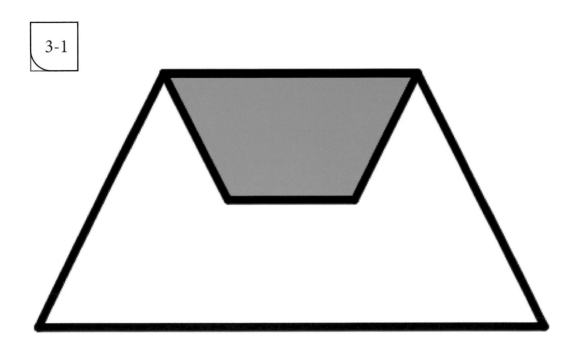

3-1

Here are two similar isosceles trapeziums/trapezoids where the small one has half the height of the big one.

What is the fraction shaded?

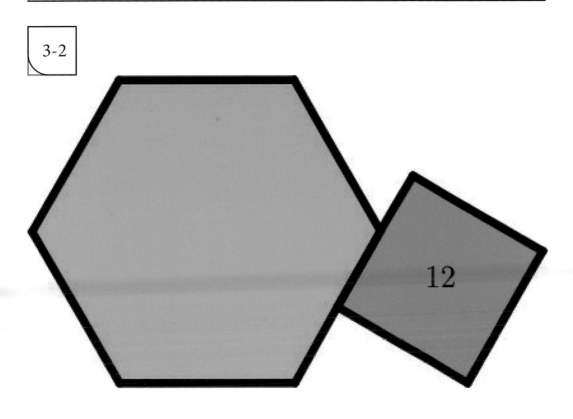

3-2

The square has area 12 and one of its sides overlaps half of the side of the regular hexagon.

What is the area of the regular hexagon?

3-3

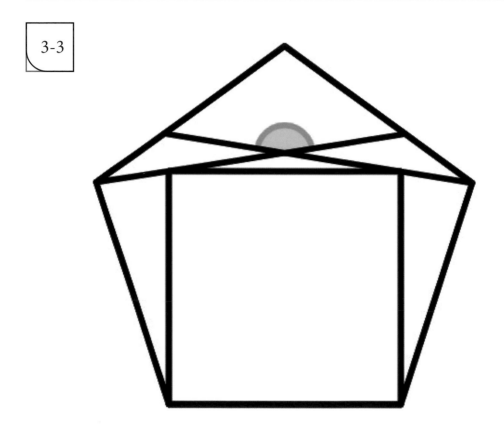

A regular pentagon and a square have the same side length.

What is the size of the angle?

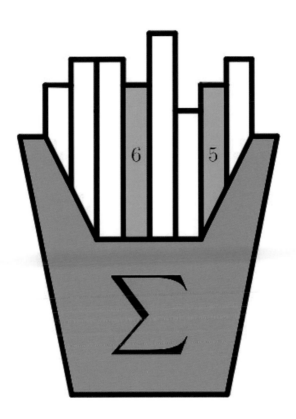

A box of fries has two shaded potatoes with areas 6 and 5. All the potatoes in the box have width 1.

What fraction of fries are shaded?

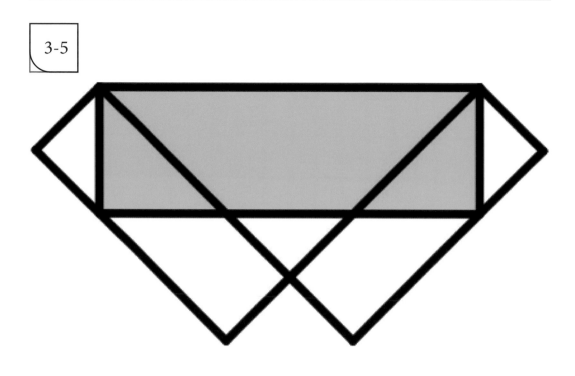

Two congruent rectangles have area 6.

What is the area of the blue rectangle?

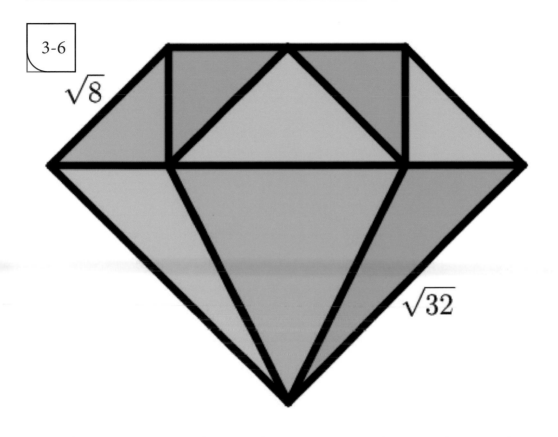

A perfectly symmetric diamond is represented above. Its bottom forms an angle of 90°.

What is its area?

3-7

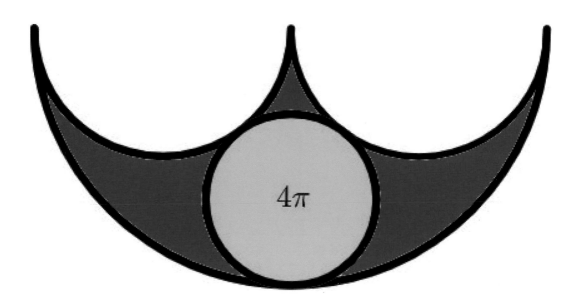

4π

The light blue circle is inscribed into a big semicircle and two small semicircles with the same radius.

What is the dark blue shaded area?

3-8

A bird is represented above.

What is the area of the bird (light blue shaded area)?

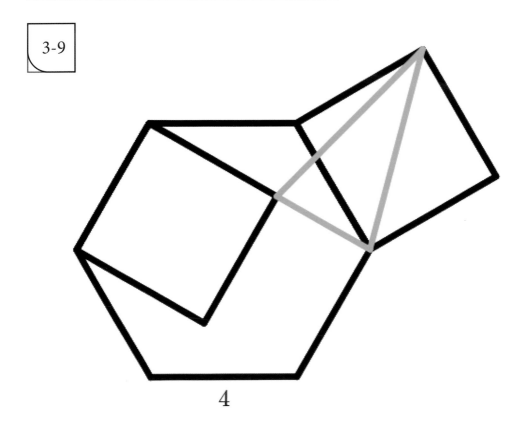

3-9

4

A regular hexagon and two squares with side length 4 are represented above.

What is the area of the isosceles triangle.

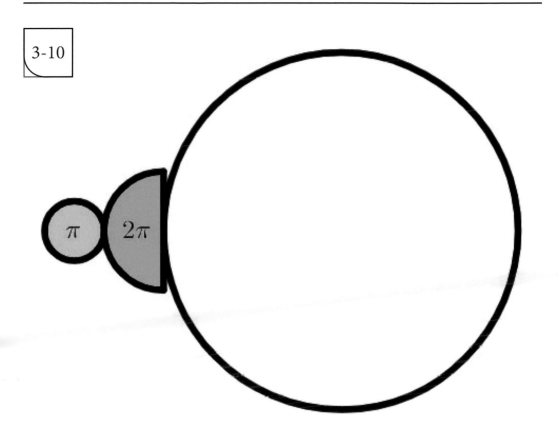

A semicircle is tangent to two circles.

What is the area of the big circle?

Solutions and Answers

3-1 Answer: $\dfrac{1}{4}$

Let a be the length of the bottom base and b be the length of the top base of the big trapezium.

So, b is also the length of the bottom base (the bigger one) of the small trapezium.

Let c be the length of the top base (the small one) of the small trapezium.

As the isosceles trapeziums are similar and the height of the small one is the half of the height of the big one, then $a = 2b$ and $b = 2c$.

Then, the area of the small trapezium is

$$\frac{b+c}{2} \times \frac{h}{2} = \frac{2c+c}{4} \times h = \frac{3ch}{4}$$

and the area of the big trapezium is

$$\frac{a+b}{2} \times h = \frac{2b+b}{2} \times h = \frac{3b}{2} \times h = \frac{6c}{2} \times h = 3ch$$

So, the fraction shaded is $\dfrac{\dfrac{3ch}{4}}{3ch} = \dfrac{1}{4}$.

3-2 Answer: $24\sqrt{3}$

We can draw in the diagram two similar right triangles.
Then,

$$\frac{\sqrt{12}}{\dfrac{a}{2}} = \frac{\dfrac{\sqrt{3}}{2}a + \sqrt{12}}{a}$$

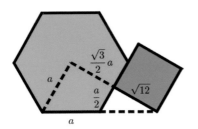

$$\frac{2\sqrt{12}}{a} = \frac{\frac{\sqrt{3}}{2}a + \sqrt{12}}{a}$$

$$2\sqrt{12} = \frac{\sqrt{3}}{2}a + \sqrt{12}$$

$$\frac{\sqrt{3}}{2}a = \sqrt{12}$$

$$\frac{\sqrt{3}}{2}a = 2\sqrt{3}$$

$$a = 4$$

So, the area of the regular hexagon is $\dfrac{3\sqrt{3}}{2}a^2 = \dfrac{3\sqrt{3}}{2}\times(4)^2 = 24\sqrt{3}$.

3-3 Answer: 162°

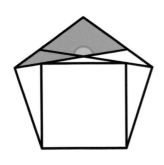

The interior angles of a regular pentagon have size 108°. Then, the small angle between the side of the regular pentagon and the side of the square is 18°.

Considering one of the two isosceles triangles adjacent to the square, each of the equal angles has size $\dfrac{180° - 18°}{2} = \dfrac{162°}{2} = 81°$. Now consider the triangle shaded:

One of the angles has size 108°, other has size $108° - 81° = 27°$

and the last one has size $180° - (108° + 27°) = 45°$.

Finally, consider the quadrilateral at the top of the regular pentagon.

So, the angle to find has size $360° - 108° - 2 \times 45° = 162°$.

3-4 Answer: $\dfrac{1}{4}$

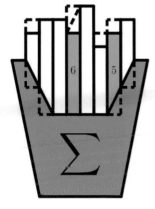

The small trapezium has bases 4 and 8.

Comparing the potato with area 6 with the potato with area 5, we get a right triangle and a square, both with area 1.

Then, there are six triangles and three squares of area 1 as we can see in the diagram to the left:

These triangles and squares are enough to find the areas of the potatoes (trapeziums and rectangles).

The sum of the areas is, from left to right:

$$3+6+7+6+8+5+5+4 - 44$$

So, the shaded potatoes are $\dfrac{6+5}{44} = \dfrac{11}{44} = \dfrac{1}{4}$ of the potatoes in the box.

3-5 Answer: 12

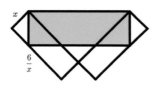

Let x and $\dfrac{6}{x}$ be the width and the length of one of the congruent rectangles with area 6.

Using the Pythagoras' theorem, we get that the width of the shaded rectangle is $\sqrt{2}x$.

The two congruent rectangles form two isosceles right triangles.

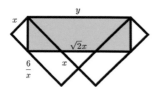

So, we can use the Thales' theorem to find the length of the shaded rectangle:

$$\frac{x}{\sqrt{2}x} = \frac{\dfrac{6}{x}}{y}$$

$$\frac{1}{\sqrt{2}} = \frac{6}{xy}$$

$$y = \frac{6\sqrt{2}}{x}$$

So, the area of the shaded rectangle is $\sqrt{2}x \times \dfrac{6\sqrt{2}}{x} = 12$.

3-6 Answer: 28

There is an isosceles right triangle with sides with length $\sqrt{32}$ in the bottom of the diamond.

Using the Pythagoras' theorem,

$$x^2 = 32 + 32$$

$$x^2 = 64$$

$$x = 8$$

In the isosceles right triangle in the top left of the diamond, using the Pythagoras' theorem again, we get that the equal sides have length 2. The same occurs for the isosceles right triangle in the top right of the diamond. Then, the line segment in the top of the diamond has length $8 - 2 \times 2 = 4$.

So, the area of the diamond is the sum of the areas of the right triangle in the bottom and the isosceles trapezium on the top:

$$\frac{\sqrt{32} \times \sqrt{32}}{2} + \frac{4+8}{2} \times 2$$

$$\frac{32}{2} + 4 + 8$$

$$16 + 12$$

$$28$$

3-7 Answer: 5π

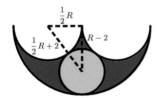

Let R be the radius of the big semicircle. Then, the radius of the two small semicircles is $\frac{1}{2}R$.

The radius of the circle is 2.

So, we can draw a right triangle on the diagram:

Using the Pythagoras' theorem,

$$\left(\frac{1}{2}R+2\right)^2 = \left(\frac{1}{2}R\right)^2 + (R-2)^2$$

$$\frac{1}{4}R^2 + 2R + 4 = \frac{1}{4}R^2 + R^2 - 4R + 4$$

$$R^2 - 6R = 0$$

$$R(R-6) = 0$$

$$R = 0 \text{ or } R = 6$$

$$R = 6$$

So, the dark blue shaded area is

$$\frac{\pi(6)^2}{2} - 2 \times \frac{\pi(3)^2}{2} - 4\pi$$

$$18\pi - 9\pi - 4\pi$$

$$5\pi$$

3-8 Answer: 69π

There are two equal semicircles that represents the wings, one semicircle that represents the head, one quarter of circle that represents the body and one quarter of circle that represents the tail.

Let R and r be the radii of the head and the tail, respectively. Then,

$$2R - r + 2R + \frac{2R - r}{2} = 17 \text{ and } R + 2R = 12$$

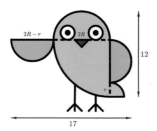

Solve the equations simultaneously, we get $R = 4$ and $R = 2$.
So, the area of the bird is

$$\pi\left(4\right)^2 + 2\times\frac{\pi\left(2\left(4\right)-2\right)^2}{2} + \frac{\pi\left(8\right)^2}{4} + \frac{\pi\left(2\right)^2}{4}$$

$$16\pi + 36\pi + 16\pi + \pi$$

$$69\pi$$

3-9 Answer: 8

As the interior angles of a regular hexagon have size $120°$, the equal angles of the isosceles triangle have size $30° + 45° = 75°$. Then, the third angle has size $180° - 2\times 75° = 180° - 150° = 30°$.

The diagonal of the square has length $4\sqrt{2}$.
So, the area of the triangle is

$$\frac{1}{2}\left(4\sqrt{2}\right)^2 \sin 30°$$

$$\frac{1}{2}\times 32\times\frac{1}{2}$$

$$\frac{32}{4}$$

$$8$$

3-10 Answer: 36π

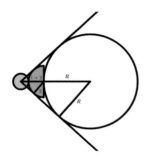

The radius of the small circle is 1 and the radius of the semicircle is 2.

We can draw two tangents to the semicircle and the big circle passing through the centre of the small circle:

As the two right triangles are similar, we have

$$\frac{R}{2} = \frac{R+3}{3}$$
$$3R = 2R + 6$$
$$R = 6$$

So, the area of the big circle is $\pi\left(6\right)^2 = 36\pi$.

Level 4

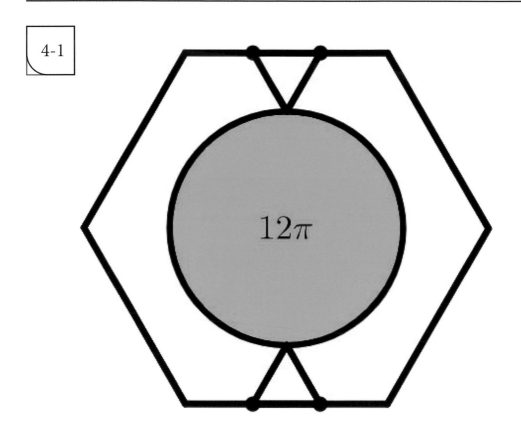

4-1

12π

A circle with area 12π and two equilateral triangles are inside a regular hexagon. The dots separate the sides in 3 equal parts.

What is the length of the side of the regular hexagon?

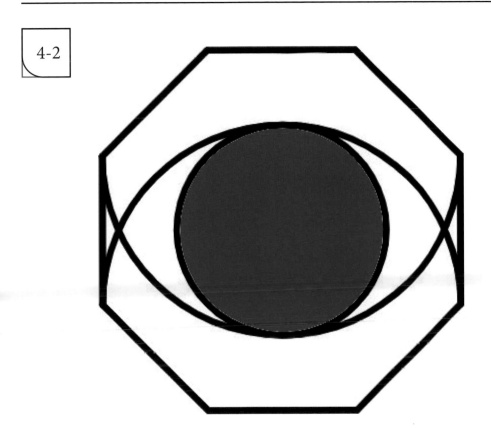

4-2

In an octagon with side with length 4, a circle is tangent on two semicircles.

What is the area of the circle?

A snail has its shell with a spiral made up with four successive semicircles with total length 20π.

What is the area of the shell?

An umbrella has the design made up with semicircles as show in the diagram above.

Given that the dark blue part has area 4π, what's the fraction of the umbrella with light blue parts?

4-5

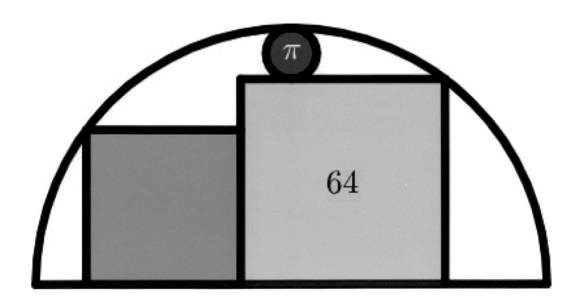

Two squares and a circle are inside a semicircle.

What is the area of the small square?

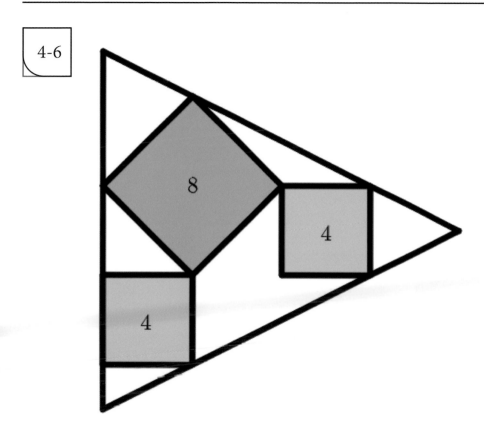

4-6

Three squares are inside an isosceles triangle.

What fraction of the triangle is shaded?

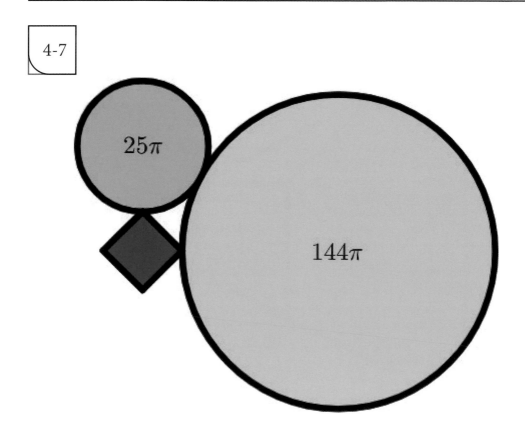

4-7

25π

144π

A square is tangent to two circles.

What is the area of the square?

4-8

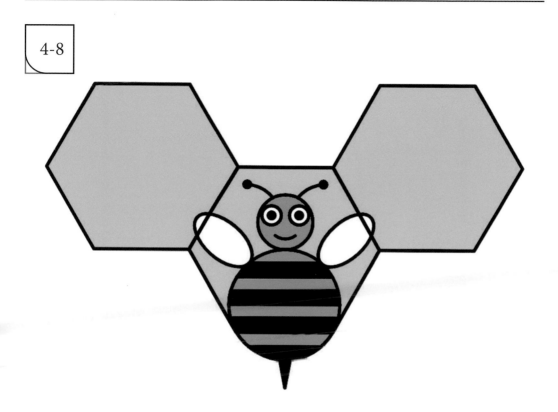

A part of a honeycomb has area 27. The blue and black stripes of the bee have the same height. The last two stripes are outside the honeycomb.

What is the area of the bee's body?

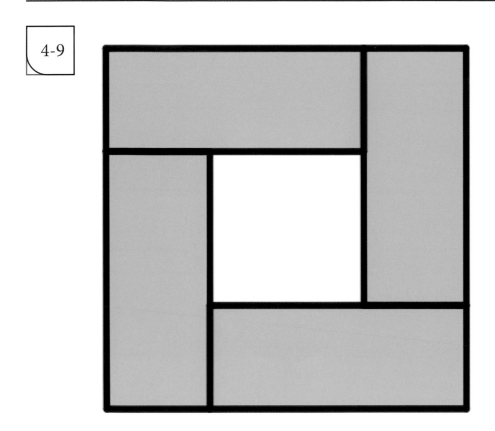

Four rectangles with area 12 form a big and a small square.

Given that the product of the sides of the two squares is 32, what is the area of the small square?

4-10

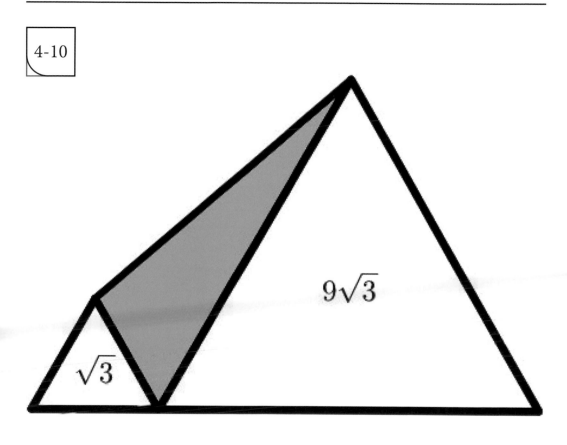

Two similar equilateral triangles are shown above.

What is the area of the shaded triangle?

Solutions and Answers

4-1 Answer: 6

As the area of the circle is 12π, its radius is

$$\pi r^2 = 12\pi$$
$$r^2 = 12$$
$$r = 2\sqrt{3}$$

Then its diameter is $4\sqrt{3}$.

The distance between two parallel sides of a regular hexagon is $\sqrt{3}l$, where l is the length of the side.

The height of each equilateral triangle is

$$\left(\frac{1}{3}l\right)^2 = \left(\frac{1}{6}l\right)^2 + h^2$$
$$\frac{1}{9}l^2 = \frac{1}{36}l^2 + h^2$$
$$h^2 - \frac{1}{9}l^2 - \frac{1}{36}l^2$$
$$h^2 = \frac{1}{12}l^2$$
$$h = \frac{1}{2\sqrt{3}}l$$

So, the length of the side of the regular hexagon is

$$4\sqrt{3} + 2h = \sqrt{3}l$$

$$4\sqrt{3} + 2 \times \frac{1}{2\sqrt{3}}l = \sqrt{3}l$$

$$4\sqrt{3} + \frac{1}{\sqrt{3}}l = \sqrt{3}l$$

$$12 + l = 3l$$

$$2l = 12$$

$$l = 6$$

4-2 Answer: 8π

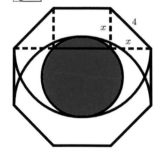

First, we need to find the radius of the two semicircles.
Let x be the height of the top trapezium.
We have a right triangle with two sides with length x and hypotenuse 4.
Using the Pythagoras' theorem,

$$4^2 = x^2 + x^2$$

$$16 = 2x^2$$

$$x^2 = 8$$

$$x = 2\sqrt{2}$$

The diameter of each semicircle is $2\sqrt{2} + 4 + 2\sqrt{2} = 4 + 4\sqrt{2}$. So, their radii are the half, i.e., $2 + 2\sqrt{2}$.

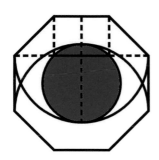

Now, we need to find the diameter of the shaded circle.

The dashed vertical line segment in the diagram has length $2\sqrt{2} + \left(2 + 2\sqrt{2}\right) = 2 + 4\sqrt{2}$.

Doing the same with the other semicircle, we get the diameter of the circle:

$$2 + 4\sqrt{2} - \left(4 + 4\sqrt{2} - \left(2 + 4\sqrt{2}\right)\right) = 4\sqrt{2}$$

Then, the radius of the circle is $2\sqrt{2}$.

So, its area is $\pi\left(2\sqrt{2}\right)^2 = 8\pi$.

4-3 Answer: $41\pi + 48$

Let's consider the diameter of the first semicircle, the biggest, equal to 4x. Then, its radius is 2x. So, its perimeter is $2x\pi$.

The diameter of the second semicircle is equal to 3x. Then, its radius is $\frac{3}{2}x$. So, its perimeter is $\frac{3}{2}x\pi$.

The diameter of the third semicircle is equal to 2x. Then, its radius is x. So, its perimeter is $x\pi$.

The diameter of the second semicircle is equal to x. Then, its radius is $\frac{1}{2}x$. So, its perimeter is $\frac{1}{2}x\pi$.

Then,

$$2x\pi + \frac{3}{2}x\pi + x\pi + \frac{1}{2}x\pi = 20\pi$$

$$\frac{4x\pi + 3x\pi + 2x\pi + x\pi}{2} = 20\pi$$

$$5x\pi = 20\pi$$

$$x = 4$$

Now, let's find out the area of the shell.

We can divide it in 3 parts: one semicircle, one quarter of circle and one trapezium as we can see in the diagram below:

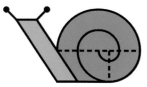

So, the area of the shell is:

$$\frac{\pi(8)^2}{2} + \frac{\pi(6)^2}{4} + \frac{6+10}{2} \times 6$$

$$41\pi + 48$$

4-4 Answer: $\frac{3}{5}$

First, let's find out the radius of the 5 small semicircles of the umbrella.

Let r be the radius of the small semicircle.

Then, the area of the dark blue part is

$$\frac{\pi(2r)^2}{2} - 2 \times \frac{\pi r^2}{2} = 4\pi$$

$$\frac{4\pi r^2}{2} - \pi r^2 = 4\pi$$

$$2\pi r^2 - \pi r^2 = 4\pi$$
$$\pi r^2 = 4\pi$$
$$r^2 = 4$$
$$r = 2$$

The total area of the umbrella is given by:

$$\frac{\pi(10)^2}{2} - 5 \times \frac{\pi(2)^2}{2}$$
$$50\pi - 10\pi$$
$$40\pi$$

The bigger light blue area has area:

$$50\pi - 2\pi - \frac{\pi(8)^2}{2}$$
$$50\pi - 2\pi - 32\pi$$
$$16\pi$$

The smaller light blue area has area:

$$\frac{\pi(6)^2}{2} - 2\pi - \frac{\pi(4)^2}{2}$$
$$18\pi - 2\pi - 8\pi$$
$$8\pi$$

So, the fraction of the umbrella with light blue parts is $\dfrac{16\pi + 8\pi}{40\pi} = \dfrac{24\pi}{40\pi} = \dfrac{3}{5}$.

4-5 Answer: 36

The radius of the circle is 1 and the length of the side of the big square is 8. Then, the radius of the semicircle is $2 \times 1 + 8 = 10$.

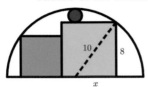

We can draw a right-angle triangle inside the big square:
Using the Pythagoras' theorem:

$$10^2 = x^2 + 8^2$$
$$100 = x^2 + 64$$
$$x^2 = 36$$
$$x = 6$$

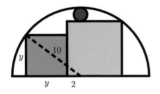

We can draw another right-angle triangle:
Using the Pythagoras' theorem again:

$$10^2 = y^2 + (y+2)^2$$
$$100 = y^2 + y^2 + 4y + 4$$
$$2y^2 + 4y - 96 = 0$$
$$y^2 + 2y - 48 = 0$$
$$y = \frac{-2 \pm \sqrt{4+192}}{2}$$
$$y = \frac{-2 \pm \sqrt{196}}{2}$$
$$y = \frac{-2 \pm 14}{2}$$

$$y = 6 \text{ or } y = -8$$

Then, $y = 6$.
So, the area of the small square is 36.

4-6 Answer: $\dfrac{1}{2}$

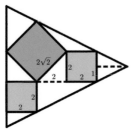

Using the area of the squares to find their side lengths and the Pythagoras' theorem, we can amend our diagram as shown.

The right triangle with sides 4 and 2 between the two smaller right triangles is similar with them, with scale factor 2.

Then, the diagram looks like this:
Calculating the hypotenuses, using the Pythagoras' theorem, of these three right triangles, we get one of the equal sides of

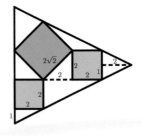

the isosceles triangle:

$$\sqrt{5} + 2\sqrt{5} + \sqrt{5}$$

$$4\sqrt{5}$$

The height of the isosceles triangle is $2+2+2+2=8$.
To find the area of the isosceles triangle we need to find its base.
Using Pythagoras' theorem again:

$$\left(4\sqrt{5}\right)^2 = x^2 + 8^2$$

$$x^2 = 80 - 64$$

$$x^2 = 16$$

$$x = 4$$

Then, the area of the isosceles triangle is $\dfrac{8 \times 8}{2} = \dfrac{64}{2} = 32$.

So, $\dfrac{4+4+8}{32} = \dfrac{16}{32} = \dfrac{1}{2}$ of the isosceles triangle is shaded.

4-7
Answer: 18

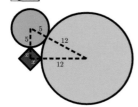

The radii of the small and big circles are 5 and 12, respectively.
Then, we get:
Using the Pythagoras' theorem:

$$\left(5+12\right)^2 = \left(5+x\right)^2 + \left(x+12\right)^2$$
$$17^2 = 25 + 10x + x^2 + x^2 + 24x + 144$$
$$289 = 2x^2 + 34x + 169$$
$$2x^2 + 34x - 120 = 0$$
$$x^2 + 17x - 60 = 0$$
$$x = \frac{-17 \pm \sqrt{289 + 240}}{2}$$
$$x = \frac{-17 \pm \sqrt{529}}{2}$$
$$x = \frac{-17 \pm 23}{2}$$
$$x = 3 \text{ or } x = -20$$
$$x = 3$$

Then, the side of the square is $3\sqrt{2}$.

So, the area of the square is $\left(3\sqrt{2}\right)^2 = 9 \times 2 = 18$.

4-8 | Answer: 4π

The area of the three regular hexagons is 27. Then, each one of them has area 9. So, the length of the side of the regular hexagon is

$$\frac{3\sqrt{3}}{2}a = 9$$

$$3\sqrt{3}a = 18$$

$$a = \frac{18}{3\sqrt{3}}$$

$$a = \frac{6}{\sqrt{3}}$$

$$a = 2\sqrt{3}$$

The bottom side of the regular hexagon where the bee lies and the diameter of the circle are represented by a white line segment: Using the Pythagoras' theorem, we can find out the radius of the circle:

$$r^2 = \left(\frac{r}{2}\right)^2 + \left(\sqrt{3}\right)^2$$

$$r^2 = \frac{r^2}{4} + 3$$

$$r^2 - \frac{r^2}{4} = 3$$

$$\frac{3r^2}{4} = 3$$

$$r^2 = 4$$
$$r = 2$$

So, the area of the bee's body is $\pi(2)^2 = 4\pi$.

4-9 Answer: 16

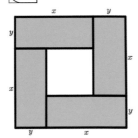

Let x and y be the length and the width of the rectangle, respectively.

Then, $xy = 12$, i.e., $x = \frac{12}{y}$

The side of the big square is $x + y$ and the side of the small square is $x - y$

So,

$$(x + y)(x - y) = 32$$
$$x^2 - y^2 = 32$$

Solving both equations simultaneously:

$$x^2 - y^2 = 32$$
$$\left(\frac{12}{y}\right)^2 - y^2 = 32$$
$$\frac{144}{y^2} - y^2 = 32$$
$$144 - y^4 = 32y^2$$
$$y^4 + 32y^2 - 144 = 0$$

Let $z = y^2$:

$$z^2 + 32z - 144 = 0$$

$$z = \frac{-32 \pm \sqrt{1024 + 576}}{2}$$

$$z = \frac{-32 \pm \sqrt{1600}}{2}$$

$$z = \frac{-32 \pm 40}{2}$$

$$z = 4$$

As $z = y^2$, $y = 2$.

Then, $x = \frac{12}{2} = 6$

So, the area of the small square is $(6-2)^2 = 4^2 = 16$

4-10 Answer: $3\sqrt{3}$

The equilateral triangles are similar. Then the scale factor is:

$$r^2 = \frac{9\sqrt{3}}{\sqrt{3}}$$

$$r^2 = 9$$

$$r = 3$$

Let's find the side length of the small equilateral triangle:

$$\frac{\sqrt{3}}{4}a^2 = \sqrt{3}$$

$$a^2 = 4$$

$$a = 2$$

Then, the side length of the other equilateral triangle is $3 \times 2 = 6$.
Using the sine rule to find the area of the shaded triangle:

$$\frac{1}{2} \times 2 \times 6 \times \sin 60°$$

$$\frac{1}{2} \times 2 \times 6 \times \frac{\sqrt{3}}{2}$$

$$3\sqrt{3}$$

Level 5

5-1

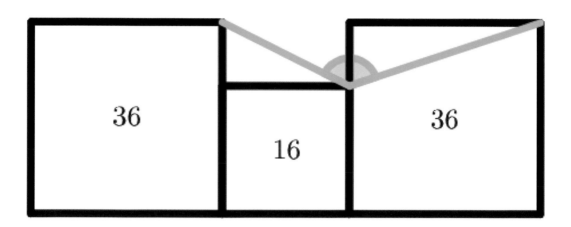

A square with area 16 is between two congruent squares with area 36.

What is the size of the angle?

5-2

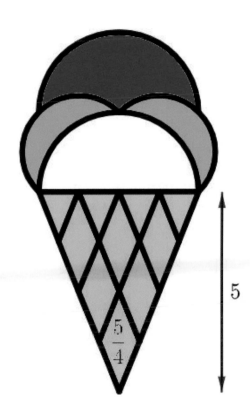

An ice-cream has four balls. The cone has height 5 and each rhombus has area $\dfrac{5}{4}$.

What is the area of the dark blue ball?

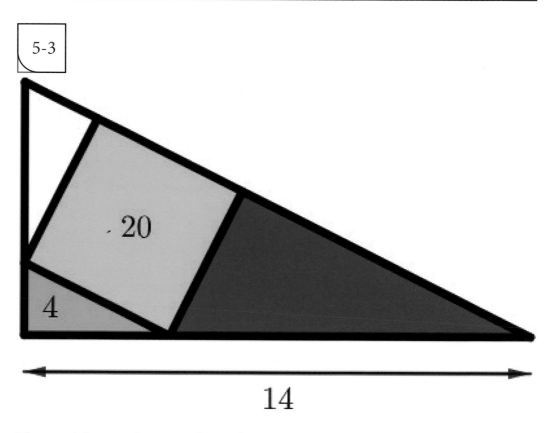

Three right-angle triangle and one square are inside of one big right-angle.

What is the area of the dark blue right-angle triangle?

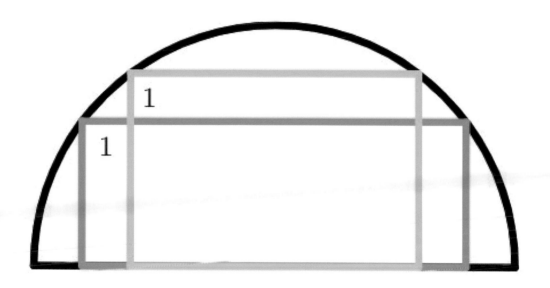

Two rectangles with area 24 are inside a semicircle.

What is the radius of the semicircle?

5-5

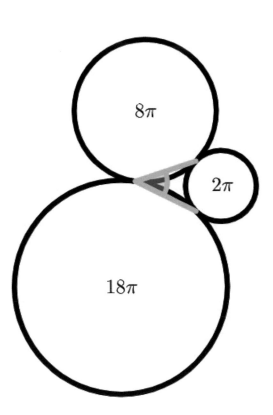

Three circles are tangent with each other.

What is the size of the angle?

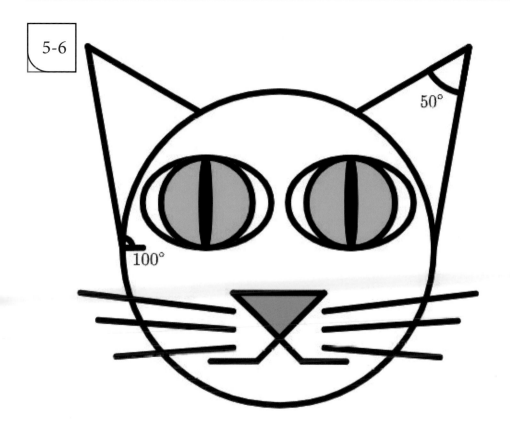

5-6

50°

100°

The diagram represents a cat's face that is perfectly symmetric. Its eyes are circles with area π.

What is the area of the cat's face?

5-7

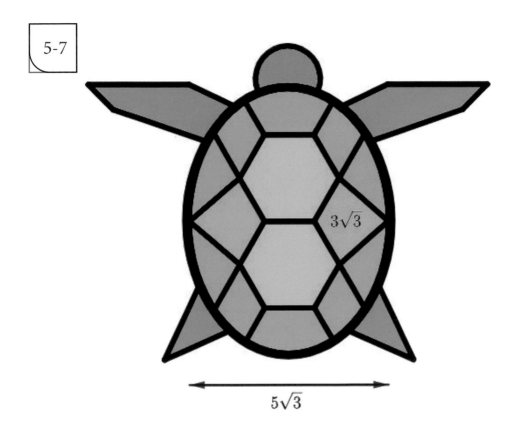

In the shell of a turtle there are two congruent hexagons and two congruent kites.

What is the area of the two hexagons?

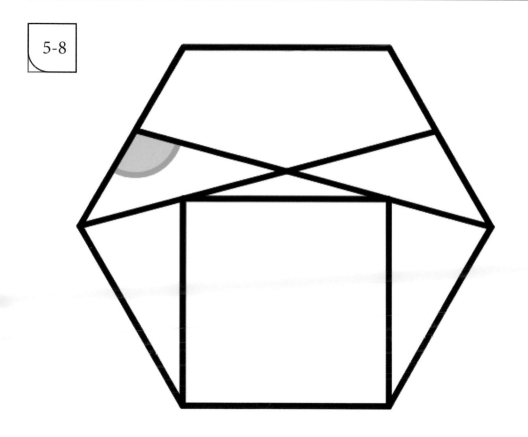

5-8

A regular hexagon and a square have the same side length.

What is the size of the angle?

A perfectly symmetric panda's face is show below. The small circle is tangent to the big circle and passes through its centre. Inside the small circle, a triangle meets the extremities of two congruent semicircles at its centre. The biggest side of the triangle has the same length of the diameter of the semicircles.

What is the radius of the panda's face?

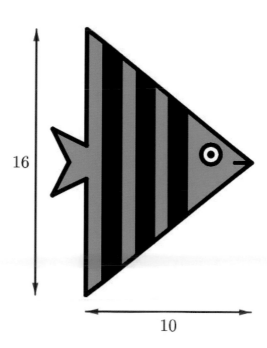

A fish is represented by an isosceles triangle as shown above. All the stripes, blue and black, have height 1, including the blue stripe next to the tail.

What is the fraction of the body of the fish (without the tail) that is painted with black?

Solutions and Answers

5-1 Answer: 135°

The square in the middle has length side 4 and the other two squares have length side 6.

Let's decompose the blue angle into two angles. We get two right-angle triangles like we see in the diagram below:

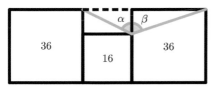

The angle in the leftmost right triangle, α, has an opposite side with length 4 and an adjacent side with side length 2. So, $\tan\alpha = \dfrac{4}{2} = 2$.

The angle in the rightmost right triangle, β, has an opposite side with length 6 and an adjacent side with side length 2. So, $\tan\beta = \dfrac{6}{2} = 3$.

Using the compound formula for the tangent:

$$\tan(\alpha + \beta) = \frac{\tan\alpha + \tan\beta}{1 - \tan\alpha\tan\beta}$$

$$\tan(\alpha + \beta) = \frac{2 + 3}{1 - 2 \times 3}$$

$$\tan(\alpha + \beta) = -1$$

As $\alpha + \beta$ is an obtuse angle, then $\alpha + \beta = 135°$.

5-2 Answer: $\dfrac{3\pi}{2} + \sqrt{2}$

Let p and q be the small and the big diagonal of the rhombus.

The area of the rhombus is $\dfrac{pq}{2} = \dfrac{5}{4}$. Then, $p = \dfrac{5}{2q}$.

The height of the isosceles triangle is 5. So, $2q = 5$ and $p = \dfrac{5}{5} = 1$.

Then, the radius of the white semicircle is 2 and its area is $\dfrac{\pi(2)^2}{2} = \dfrac{4\pi}{2} = 2\pi$.

We can draw the diameters of the semicircles of the balls in light blue and dark blue as follows:

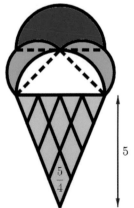

The radius of the semicircle that represents the ball in dark blue is the same of the radius of the semicircle that represents the ball in white.

The diameter of the two small semicircles that represent the balls in light blue is $2\sqrt{2}$. Then, their radius is $\sqrt{2}$.

Using the formula for the area of a segment of a circle, the area of both segments is:

$$2 \times \left(\dfrac{1}{2} \times \left(\sqrt{2} \right)^2 \left(\dfrac{\pi}{4} - \sin\left(\dfrac{\pi}{4} \right) \right) \right)$$

$$2 \left(\dfrac{\pi}{4} - \dfrac{\sqrt{2}}{2} \right)$$

$$\frac{\pi}{2} - \sqrt{2}$$

So, the area of the dark blue ball is $2\pi - \left(\frac{\pi}{2} - \sqrt{2}\right) = 2\pi - \frac{\pi}{2} + \sqrt{2} = \frac{3\pi}{2} + \sqrt{2}$.

5-3 Answer: 20

Let x and y be the base and the height, respectively, of the right triangle of area 4. The side length of the square is $\sqrt{20}$.

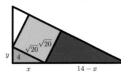

Then, we get our diagram like this:

The light blue and dark blue triangles are similar.

Then,

$$\frac{y}{\sqrt{20}} = \frac{\sqrt{20}}{14 - x}$$

$$y = \frac{20}{14 - x}$$

The area of the light blue triangle is 4. Then,

$$\frac{xy}{2} = 4$$

$$xy = 8$$

$$\frac{20x}{14 - x} = 8$$

$$20x = 112 - 8x$$

$$28x = 112$$

$$x = 4$$

So, the hypotenuse of the dark blue triangle is $14 - 4 = 10$.

The scale factor between the dark blue and the light blue triangles is $\dfrac{10}{\sqrt{20}} = \dfrac{10}{2\sqrt{5}} = \dfrac{5}{\sqrt{5}}$.

Then, the ration of their areas is $\left(\dfrac{5}{\sqrt{5}}\right)^2 = \dfrac{25}{5} = 5$.

So, the area of the dark blue triangle is $5 \times 4 = 20$.

| 5-4 | Answer: 5

Let x and y be, respectively, the length and the width of the light blue rectangle.

Then, $xy = 24$.

The length and the width of the dark blue rectangle are $x+2$ and $y-1$.
Then,

$$(x+2)(y-1) = 24$$
$$xy - x + 2y - 2 = 24$$

So,

$$xy = xy - x + 2y - 2$$
$$-x + 2y - 2 = 0$$
$$x = 2y - 2$$

As $xy = 24$, $y = \dfrac{24}{x}$.

Then, substituting the value of y in the first equation:

$$x = 2\left(\frac{24}{x}\right) - 2$$

$$x^2 = 48 - 2x$$

$$x^2 + 2x - 48 = 0$$

$$x = \frac{-2 \pm \sqrt{4 + 192}}{2}$$

$$x = \frac{-2 \pm \sqrt{196}}{2}$$

$$x = \frac{-2 \pm 14}{2}$$

$$x = 6 \ \text{ or } \ x = -8$$

$$x = 6$$

Then, $y = \dfrac{24}{6} = 4$.

Let R be the radius of the semicircle. Then, using the Pythagoras' theorem:

$$R^2 = \left(\frac{6}{2}\right)^2 + 4^2$$

$$R^2 = 3^2 + 4^2$$

$$R^2 = 9 + 16$$

$$R^2 = 25$$

$$R = 5$$

5-5 Answer: 45°

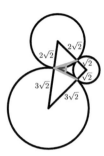

The radii of the circles are $\sqrt{2}$, $2\sqrt{2}$ and $3\sqrt{2}$.

Representing them in our diagram, we get a right triangle with sides with lengths $3\sqrt{2}$, $4\sqrt{2}$ and $5\sqrt{2}$:

A circle with radius $\sqrt{2}$ passes through the three tangent points, called incircle.

Then, the angle has size equal to the size of the angle formed by the in-centre (centre of the incircle) and the tangent points in the two smallest circles, that is 90°.

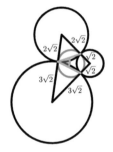

So, the size of the angle is $\dfrac{90°}{2} = 45°$.

5-6 Answer: 12π

We can draw a concave pentagon as shown to the left:

As the cat's face is symmetric, we get 2 congruent scalene triangles where one of their sides is the diameter of the circle.

Considering one of them, its angles have size 50°, 100° and

$$180° - \left(50° + 100°\right) = 30°.$$

Let's consider the isosceles triangle. Their angles have size $30°$, $30°$ and $180° - 2 \times 30° = 120°$.

As the radius of the eyes are 1, the triangle has height 2.

Then, the radius of the circle that represents the cat's face is

$$2\tan 60°$$

$$2\sqrt{3}$$

So, the area of the circle is $\pi\left(2\sqrt{3}\right)^2 = 12\pi$.

5-7 | Answer: 9

The area of the kite is $3\sqrt{3}$.

Then,

$$\frac{p \times q}{2} = 3\sqrt{3}$$

$$p \times q = 6\sqrt{3} \quad (1)$$

where p and q are the diagonals of the kite.

The angle formed between the two congruent hexagons has size $120°$, as the internal

angles of a regular hexagon have size $120°$ and $\dfrac{360°}{3} = 120°$.

Let p and q be the small and the big diagonals of the kite, respectively. Let a be the side length of the regular hexagon.

Then,

$$\frac{p}{2} = a\sin 60°$$

$$\frac{p}{2} = a \times \frac{\sqrt{3}}{2}$$

$$p = \sqrt{3}a$$

As the width of the shell of the turtle is $5\sqrt{3}$,

$$2q + a = 5\sqrt{3}$$

$$2q = 5\sqrt{3} - a$$

$$q = \frac{5\sqrt{3} - a}{2}$$

Substituting the values of p and q into the equation (1), we get:

$$\sqrt{3}a \times \frac{5\sqrt{3} - a}{2} = 6\sqrt{3}$$

$$15a - \sqrt{3}a^2 = 12\sqrt{3}$$

$$\sqrt{3}a^2 - 15a + 12\sqrt{3} = 0$$

$$a = \frac{15 \pm \sqrt{225 - 144}}{2\sqrt{3}}$$

$$a = \frac{15 \pm \sqrt{81}}{2\sqrt{3}}$$

$$a = \frac{15 \pm 9}{2\sqrt{3}}$$

$$a = \frac{24}{2\sqrt{3}} \text{ or } a = \frac{6}{2\sqrt{3}}$$

$$a = 4\sqrt{3} \text{ or } a = \sqrt{3}$$

As p is the small diagonal and q the big diagonal of the kite, $a = \sqrt{3}$.
So, the area of the 2 regular hexagons is

$$2 \times \left(\frac{3\sqrt{3}}{2} \times \sqrt{3} \right)$$

9

5-8 Answer: 105°

The interior angles of a regular hexagon have size $120°$. Then, the small angle between the side of the regular pentagon and the side of the square is $30°$.
Considering one of the two isosceles triangles adjacent to the square, each of the equal angles has size $\dfrac{180° - 30°}{2} = 75°$.

As the interior angles of a regular hexagon have size $120°$, the adjacent angle has size $120° - 75° = 45°$.
Consider the triangle shaded:
The objective is to find its other acute angle.
Now, consider the isosceles triangle on the top of the square.

Their acute angles have size $180° - 75° - 90° = 15°$. Then, its obtuse angle has size $180° - 2 \times 15° = 150°$.

Then, the other acute angle of the shaded triangle has size $180° - 150° = 30°$.

So, the angle we are looking for has size $180° - 45° - 30° = 105°$.

| 5-9 | Answer: 8 |

Let b and h be the base and the height of the triangle, respectively.
Then,

$$\frac{b \times h}{2} = 4$$

$$b \times h = 8$$

$$b = \frac{8}{h}$$

The radius of the small circle is $h + 2$ that is also the diameter of each semicircle that is b. Then, $b = h + 2$.

So,

$$h + 2 = \frac{8}{h}$$

$$h^2 + 2h = 8$$

$$h^2 + 2h - 8 = 0$$

$$h = \frac{-2 \pm \sqrt{4 + 32}}{2}$$

$$h = \frac{-2 \pm \sqrt{36}}{2}$$

$$h = \frac{-2 \pm 6}{2}$$

$$h = \frac{4}{2} \text{ or } h = -\frac{8}{2}$$

$$h = 2$$

Then, $b = 2 + 2 = 4$.

So, the radius of the big circle is $2 \times 4 = 8$.

5-10 Answer: $\dfrac{39}{100}$

Using the Thales' theorem, we get successively:

- $\dfrac{16}{10} = \dfrac{x}{9} \Rightarrow x = 14.4$

- $\dfrac{16}{10} = \dfrac{x}{8} \Rightarrow x = 12.8$

- $\dfrac{16}{10} - \dfrac{x}{7} \Rightarrow x = 11.2$

- $\dfrac{16}{10} = \dfrac{x}{6} \Rightarrow x = 9.6$

- $\dfrac{16}{10} = \dfrac{x}{5} \Rightarrow x = 8$

- $\dfrac{16}{10} = \dfrac{x}{4} \Rightarrow x = 6.4$

Then, the areas of the three black trapeziums are:

- $A_1 = \dfrac{14.4 + 12.8}{2} \times 1 = 13.6$

- $A_2 = \dfrac{11.2 + 9.6}{2} \times 1 = 10.4$

- $A_3 = \dfrac{8 + 6.4}{2} \times 1 = 7.2$

The total area of the trapeziums is $13.6 + 10.4 + 7.2 = 31.2$.

The area of the isosceles triangle is $\dfrac{16 \times 10}{2} = 80$.

So, the fraction of the body of the fish (without the tail) is painted with black is

$$\frac{31.2}{80} = \frac{39}{100}.$$

Level 6

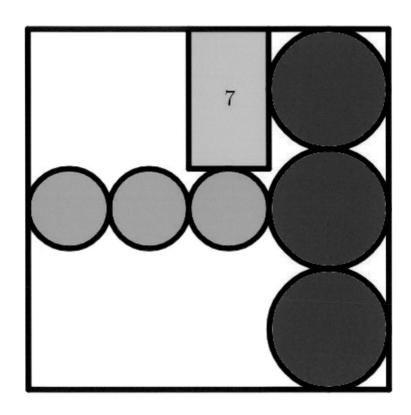

Six circles and a rectangle with area 7 are inside a square.

What is the area of the square?

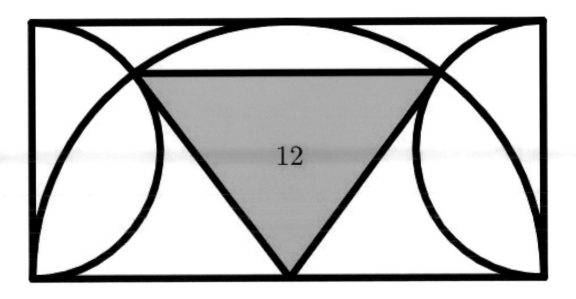

An isosceles triangle with area 12 and three semicircles are inside a rectangle.

What is the area of the rectangle?

6-3

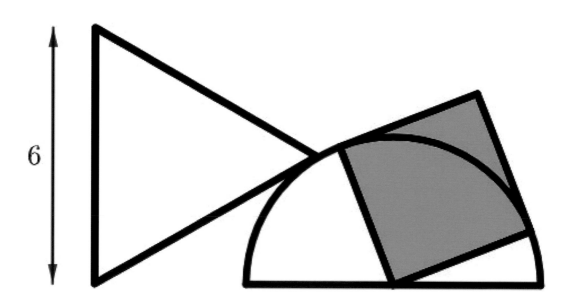

6

One of the vertices of an equilateral triangle is tangent to a semicircle. A square has two of its vertices on the semicircle and one of them lies on the centre of the semicircle.

What is the area of the square?

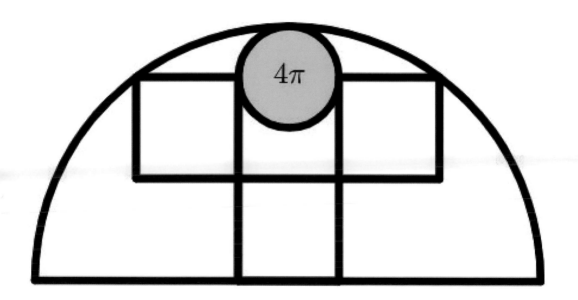

Three congruent squares and a circle are inside a semicircle.

What fraction of the semicircle is shaded?

6-5

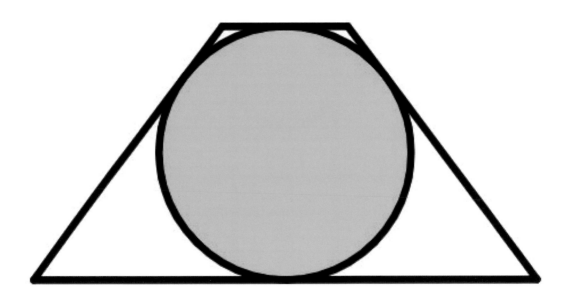

A circle is inscribed in an isosceles trapezium that has perimeter and area 20.

What is the area of the circle?

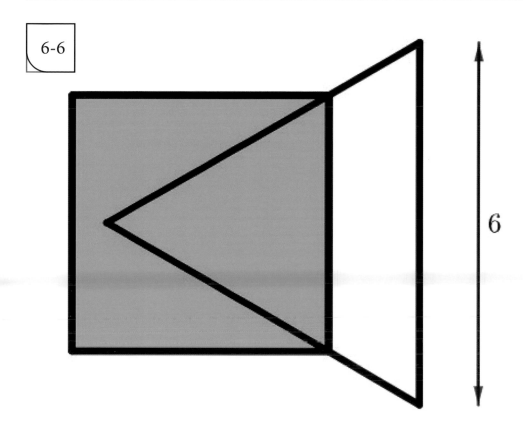

6-6

The small equilateral triangle and the isosceles trapezium have the same area.

What is the area of the shaded square?

6-7

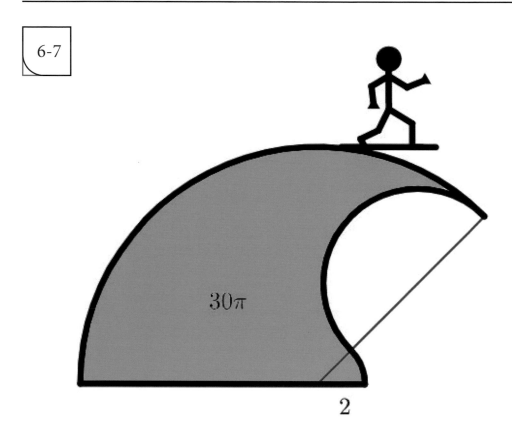

30π

2

A wave with area 30π is represented above, where the radius of the octant is 2.

What is the height of the wave?

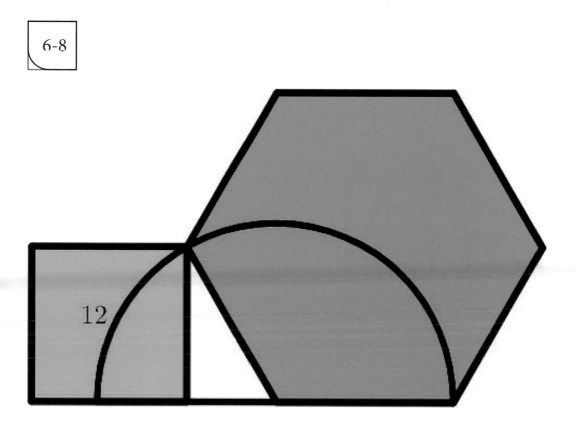

The diagram shows a square with area 12 and a regular hexagon.

What is the area of the semicircle.

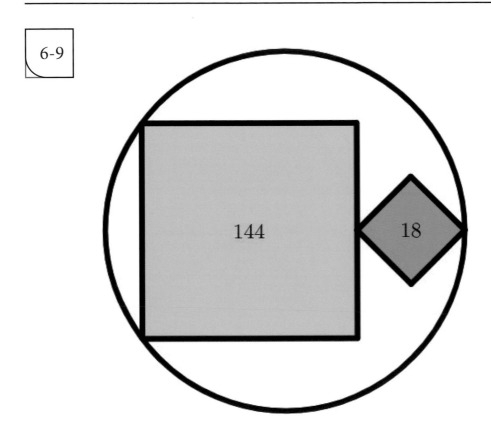

6-9

144

18

Two squares meet inside a circle.

What's the area of the circle?

6-10

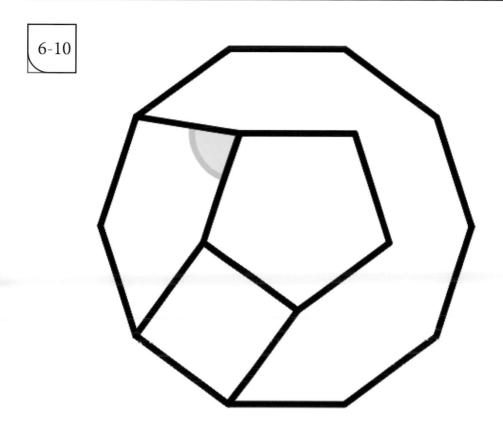

A square and a regular pentagon are inside a regular decagon.

What is the size of the angle?

Solutions and Answers

6-1 Answer: 81

Let x and y be the radii of the small and big circles, respectively.

Then, $6x + 2y = 6y$. So, $x = \dfrac{2}{3}y$.

As the area of the rectangle is 7,

$$2x(3y - x) = 7$$

$$6xy - 2x^2 = 7$$

$$6 \times \frac{2}{3} y \times y - 2 \times \left(\frac{2}{3}y\right)^2 = 7$$

$$4y^2 - 2 \times \frac{4}{9} y^2 = 7$$

$$4y^2 - \frac{8}{9} y^2 = 7$$

$$28y^2 = 63$$

$$y^2 = \frac{63}{28}$$

$$y^2 = \frac{9}{4}$$

$$y = \frac{3}{2}$$

So, the area of the big square is $\left(6 \times \dfrac{3}{2}\right)^2 = \left(9\right)^2 = 81$.

6-2 | Answer: 50

Let r be the radius of the smaller semicircle. Then $2r$ is the radius of the bigger semicircle. Then, the area of the rectangle is $4r \times 2r = 8r^2$.

Consider the right-angle triangle at the bottom of the rectangle:

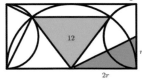

Its hypotenuse is:

$$x^2 = (2r)^2 + r^2$$
$$x^2 = 4r^2 + r^2$$
$$x^2 = 5r^2$$
$$x = \sqrt{5}r$$

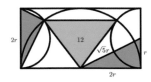

Let's consider the right-angle triangle at the left side of the rectangle:

The two right-angle are similar.

The scale factor is

$$k = \frac{2r}{\sqrt{5}r}$$
$$k = \frac{2}{\sqrt{5}}$$
$$k = \frac{2\sqrt{5}}{5}$$

Let x and y be the bigger and the smaller adjacent sides of the right angle in this triangle.

So, using the scale factor:

$$x = 2r \times \frac{2\sqrt{5}}{5}$$

$$x = \frac{4\sqrt{5}}{5}r$$

and

$$y = r \times \frac{2\sqrt{5}}{5}$$

$$y = \frac{2\sqrt{5}}{5}r$$

Now, let's consider the small right triangle inside the right triangle at the left side of the rectangle:

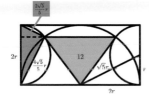

These two right-angle are also similar.

Using Thales' theorem to find the length of the dashed side:

$$\frac{2r}{\frac{2\sqrt{5}}{5}r} = \frac{\frac{4\sqrt{5}}{5}r}{z}$$

$$z = \frac{4\sqrt{5}}{5}r \times \frac{\sqrt{5}}{5}$$

$$z = \frac{4}{5}r$$

To find the smaller side of the smaller right triangle, t, we use the Thales' theorem again:

$$\frac{2r}{\frac{2\sqrt{5}}{5}r} = \frac{\frac{2\sqrt{5}}{5}r}{t}$$

$$t = \frac{2\sqrt{5}}{5}r \times \frac{\sqrt{5}}{5}$$

$$t = \frac{2}{5}r$$

Finally, let's find the base and the height of the given isosceles triangle. The base is:

$$b = 4r - 2 \times \frac{4}{5}r$$

$$b = 4r - \frac{8}{5}r$$

$$b = \frac{12}{5}r$$

The height is:

$$h = 2r - \frac{2}{5}r$$

$$h = \frac{8}{5}r$$

Then, as the area of the isosceles triangle is 12:

$$\frac{\frac{12}{5}r \times \frac{8}{5}r}{2} = 12$$

$$\frac{12}{5}r \times \frac{8}{5}r = 24$$

$$\frac{96}{25}r^2 = 24$$

$$r^2 = 24 \times \frac{25}{96}$$

$$r^2 = \frac{25}{4}$$

So, the area of the rectangle is $8 \times \frac{25}{4} = 50$.

6-3 Answer: 12

Let consider the two similar right-angle triangles in light blue, both have one of its sides the radius of the semicircle.

The two right-angle triangle have interior angles $30°$, $60°$ and $90°$.

The height of both right-angle triangle (dashed line segment) is

$$6 \sin 30° = 6 \times \frac{1}{2} = 3.$$

Part of the hypotenuse of the big right-angle triangle (the one that does not belong

to the small right-angle triangle) have length $6\cos 30° = 6 \times \dfrac{\sqrt{3}}{2} = 3\sqrt{3}$.

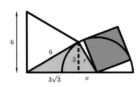

Let r be the radius of the semicircle and x be the other part of the hypotenuse of the big right-angle triangle.
We have:

$$\frac{r}{x} = \frac{6}{3}$$

$$\frac{r}{x} = 2$$

$$r = 2x$$

Also,

$$\frac{3\sqrt{3}+x}{r} = \frac{r}{x}$$

$$\frac{3\sqrt{3}+x}{2x} = \frac{2x}{x}$$

$$\frac{3\sqrt{3}+x}{2x} = 2$$

$$3\sqrt{3}+x = 4x$$

$$3x = 3\sqrt{3}$$

$$x = \sqrt{3}$$

Then,

$$r = 2\sqrt{3}$$

So, the area of the square is $\left(2\sqrt{3}\right)^2 = 12$.

6-4 Answer: $\dfrac{2}{25}$

Let x be the side of the square and r be the radius of the semicircle.

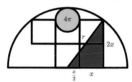

Consider the right-angle triangle:

Using Pythagoras' theorem:

$$r^2 = \left(2x\right)^2 + \left(\frac{3}{2}x\right)^2$$

$$r^2 = 4x^2 + \frac{9}{4}x^2$$

$$r^2 = \frac{25}{4}x^2$$

$$r = \frac{5}{2}x$$

When we consider the radius of the semicircle and the radius of the circle (that is 2) in the same line, i.e., a vertical line:

$$r = 2x + 2$$

Then,

$$\frac{5}{2}x = 2x + 2$$

$$5x = 4x + 4$$

$$x = 4$$

And $r = 2 \times 4 + 2 = 10$.

The area of the semicircle is $\dfrac{\pi\left(10\right)^2}{2} = 50\pi$.

So, the fraction of the semicircle that is shaded is $\dfrac{4\pi}{50\pi} = \dfrac{2}{25}$.

6-5 Answer: 4π

Let r be the radius of the circle, x be the line segment between the point of tangency and the top left/right vertex and y be the line segment between the point of tangency and the bottom left/right vertex.

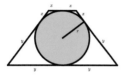

As the sides of the isosceles trapezium are tangent to the circle, we have that the perimeter is equal to

$$4x + 4y = 20$$
$$x + y = 5 \tag{1}$$

and the area is equal to

$$\frac{2x + 2y}{2} \times 2r = 20$$
$$(2x + 2y)r = 20$$
$$(x + y)r = 10 \tag{2}$$

Substituting the value of $x + y$ given by the equation (1) into the equation (2):

$$5r = 10$$
$$r = 2$$

So, the area of the circle is $\pi(2)^2 = 4\pi$.

6-6 Answer: 18

Let label our diagram with the letters x, y and z, respectively half of the base of the small equilateral triangle, height of the small equilateral triangle and height of the isosceles trapezium.

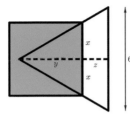

Using the Pythagoras' theorem in the big equilateral triangle:

$$6^2 = (y+z)^2 + 3^2$$
$$(y+z)^2 = 36-9$$
$$(y+z)^2 = 27$$
$$y+z = 3\sqrt{3}$$
$$z = 3\sqrt{3} - y \qquad (1)$$

Using Thales' theorem in the half of the equilateral triangles (right triangles):

$$\frac{y+z}{3} = \frac{y}{x}$$
$$\frac{3\sqrt{3}}{3} = \frac{y}{x}$$
$$\sqrt{3} = \frac{y}{x}$$
$$x = \frac{\sqrt{3}}{3} y \qquad (2)$$

As the areas of the small equilateral triangle and the isosceles trapezium are equal:

$$\frac{2xy}{2} = \frac{6+2x}{2} \times z$$
$$xy = (3+x)z \qquad (3)$$

Substituting (1) and (2) into (3), we get:

$$\left(\frac{\sqrt{3}}{3}y\right) \times y = \left(3 + \frac{\sqrt{3}}{3}y\right) \times \left(3\sqrt{3} - y\right)$$

$$\frac{\sqrt{3}}{3}y^2 = 9\sqrt{3} - 3y + 3y - \frac{\sqrt{3}}{3}y^2$$

$$\frac{2\sqrt{3}}{3}y^2 = 9\sqrt{3}$$

$$y^2 = \frac{27}{2}$$

$$y = 3\sqrt{\frac{3}{2}}$$

Then, using (2) we get:

$$x = \frac{\sqrt{3}}{3} \times 3\sqrt{\frac{3}{2}}$$

$$x = \frac{3}{\sqrt{2}}$$

$$x = \frac{3\sqrt{2}}{2}$$

So, the area of the shaded square is $\left(2 \times \dfrac{3\sqrt{2}}{2}\right)^2 = \left(3\sqrt{2}\right)^2 = 18$.

6-7 Answer: 10

Let r be the radius of the big sector of the circle.
Then,

$$\frac{3\pi}{8}r^2 + \frac{\pi}{8} \times 2^2 - \frac{\pi}{2}\left(\frac{r-2}{2}\right)^2 = 30\pi$$

$$\frac{3\pi}{8}r^2 + \frac{4\pi}{8} - \frac{\pi}{2}\left(\frac{r^2 - 4r + 4}{4}\right) = 30\pi$$

$$\frac{3\pi}{8}r^2 + \frac{4\pi}{8} - \frac{\pi r^2 - 4\pi r + 4\pi}{8} = 30\pi$$

$$3\pi r^2 + 4\pi - \pi r^2 + 4\pi r - 4\pi = 240\pi$$

$$2\pi r^2 + 4\pi r - 240\pi = 0$$

$$r^2 + 2r - 120\pi = 0$$

$$(r+12)(r-10) = 0$$

$$r = -12 \text{ or } r = 10$$

$$r = 10$$

So, the height of the wave is 10.

6-8 Answer: 8π

Let r be the radius of the semicircle. It is also the side of the regular hexagon.

The square has side length $\sqrt{12}$.

As the interior angles of the regular hexagon have size $120°$, the white right triangle has angles with sizes $30°$, $60°$ and $90°$.

Let x be the small side of this triangle. Then,

$$\sin 30° = \frac{x}{r}$$

$$\frac{1}{2} = \frac{x}{r}$$

$$x = \frac{1}{2}r$$

Using the Pythagoras' theorem:

$$r^2 = \left(\sqrt{12}\right)^2 + x^2$$

$$r^2 = 12 + \left(\frac{1}{2}r\right)^2$$

$$r^2 = 12 + \frac{1}{4}r^2$$

$$\frac{3}{4}r^2 = 12$$

$$r^2 = 12 \times \frac{4}{3}$$

$$r^2 = 16$$

$$r = 4$$

So, the area of the semicircle is $\dfrac{\pi\left(4\right)^2}{2} = 8\pi$.

Answer: 100π

Let r be the radius of the circle.

The side length of the big square is 12 and of the small square is $\sqrt{18}$.
Let x be the distance between the centre of the circle and the point of tangency between the two squares.

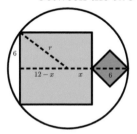

We get our diagram as follows:
So,

$$r = x + 6$$

and

$$r^2 = 6^2 + (12 - x)^2$$

Then,

$$(x+6)^2 = 6^2 + (12-x)^2$$
$$x^2 + 12x + 36 = 36 + 144 - 24x + x^2$$
$$36x = 144$$
$$x - 4$$

The radius of the circle $4 + 6 = 10$.

So, the area of the circle is $\pi(10)^2 = 100\pi$.

6-10 Answer: 81°

First, let's consider the isosceles triangle adjacent to the square:

The size of each interior angle of a regular decagon is 144°.

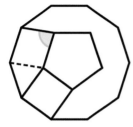

Then, the size of the unequal size is $144° - 90° = 54°$. So, the two equal angles have size $\dfrac{180° - 54°}{2} = 63°$.

As the sizes of the interior angles of a square and a regular pentagon are 90° and 108°, the angle that shares the same side of the quadrilateral is $360° - 90° - 108° - 63° = 99°$.

Using parallel straight lines and the sizes of the angles, we see that the blue sides of the quadrilateral are parallel as they form the same angle with the vertical straight lines.

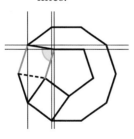

The quadrilateral has one pair of opposite sides both congruent and parallel so, it is a parallelogram.

As one of the angles of the parallelogram has size 99°, the asked angle has size $180° - 99° = 81°$.